ENERGY SECTOR STANDARD
OF THE PEOPLE'S REPUBLIC OF CHINA

中华人民共和国能源行业标准

Code for Design of AC 110 kV~500 kV Power Cable Systems for Hydropower Station

水力发电厂交流 110 kV~500 kV 电力电缆工程设计规范

NB/T 10498-2021

Replace DL/T 5228-2005

Chief Development Department: China Renewable Energy Engineering Institute
Approval Department: National Energy Administration of the People's Republic of China
Implementation Date: July 1, 2021

China Water & Power Press

中国水利水电出版社

Beijing 2024

All rights reserved. No part of this publication may be reproduced, stored in a retrieval system, or transmitted in any form or by any means—electronic, mechanical, photocopying, recording or otherwise, without prior written permission of the publisher.

图书在版编目（CIP）数据

水力发电厂交流110 kV~500 kV电力电缆工程设计规范：NB/T 10498-2021 = Code for Design of AC 110 kV~500 kV Power Cable Systems for Hydropower Station (NB/T 10498-2021)：英文 / 国家能源局发布. 北京：中国水利水电出版社，2024. 8. -- ISBN 978-7-5226-2644-4

Ⅰ. TV73-65

中国国家版本馆CIP数据核字第20241QU032号

ENERGY SECTOR STANDARD
OF THE PEOPLE'S REPUBLIC OF CHINA
中华人民共和国能源行业标准

Code for Design of AC 110 kV~500 kV Power Cable
Systems for Hydropower Station
水力发电厂交流110 kV~500 kV电力电缆
工程设计规范
NB/T 10498-2021
Replace DL/T 5228-2005
（英文版）

Issued by National Energy Administration of the People's Republic of China
国家能源局　发布
Translation organized by China Renewable Energy Engineering Institute
水电水利规划设计总院　组织翻译
Published by China Water & Power Press
中国水利水电出版社　出版发行
　　Tel: (+ 86 10) 68545888　68545874
　　sales@mwr.gov.cn
　　Account name: China Water & Power Press
　　Address: No.1, Yuyuantan Nanlu, Haidian District, Beijing 100038, China
　　http: //www.waterpub.com.cn
中国水利水电出版社微机排版中心　排版
北京中献拓方科技发展有限公司　印刷
184mm×260mm　16开本　3印张　95千字
2024年8月第1版　2024年8月第1次印刷
Price（定价）：￥500.00

Introduction

This English version is one of China's energy sector standard series in English. Its translation was organized by China Renewable Energy Engineering Institute authorized by National Energy Administration of the People's Republic of China in compliance with relevant procedures and stipulations. This English version was issued by National Energy Administration of the People's Republic of China in Announcement [2023] No. 5, dated October 11, 2023.

This version was translated from the Chinese Standard NB/T 10498-2021, *Code for Design of AC 110 kV~500 kV Power Cable Systems for Hydropower Station*, published by China Water & Power Press. The copyright is reserved by National Energy Administration of the People's Republic of China. In the event of any discrepancy in the implementation, the Chinese version shall prevail.

Many thanks go to the staff from the relevant standard development organizations and those who have provided generous assistance in the translation and review process.

For further improvement of the English version, any comments and suggestions are welcome and should be addressed to:

China Renewable Energy Engineering Institute
No. 2 Beixiaojie, Liupukang, Xicheng District, Beijing 100120, China
Website: www.creei.cn

Translating organization:

POWERCHINA Chengdu Engineering Corporation Limited

Translating staff:

YE Xiuqi   XUE Kaidan   LI Qiang   MU Kun

LI Yong   HUANG Yangkan   HE Yuanjian

Review panel members:

LIU Xiaofen	POWERCHINA Zhongnan Engineering Corporation Limited
GUO Jie	POWERCHINA Beijing Engineering Corporation Limited
CHEN Gang	POWERCHINA Huadong Engineering Corporation Limited
LI Zhongjie	POWERCHINA Northwest Engineering Corporation Limited

ZHANG Ming	Tsinghua University
HOU Yujing	China Institute of Water Resources and Hydropower Research
QIE Chunsheng	Senior English Translator
LIANG Hongli	Shanghai Investigation, Design and Research Institute Co., Ltd.

National Energy Administration of the People's Republic of China

翻译出版说明

本译本为国家能源局委托水电水利规划设计总院按照有关程序和规定，统一组织翻译的能源行业标准英文版系列译本之一。2023年10月11日，国家能源局以2023年第5号公告予以公布。

本译本是根据中国水利水电出版社出版的《水力发电厂交流110 kV~500 kV 电力电缆工程设计规范》NB/T 10498—2021翻译的，著作权归国家能源局所有。在使用过程中，如出现异议，以中文版为准。

本译本在翻译和审核过程中，本标准编制单位及编制组有关成员给予了积极协助。

为不断提高本译本的质量，欢迎使用者提出意见和建议，并反馈给水电水利规划设计总院。

地址：北京市西城区六铺炕北小街2号
邮编：100120
网址：www.creei.cn

本译本翻译单位：中国电建集团成都勘测设计研究院有限公司

本译本翻译人员：叶修齐　薛凯丹　李　强　穆　焜
　　　　　　　　李　勇　黄扬堪　何远健

本译本审核人员：

　刘小芬　中国电建集团中南勘测设计研究院有限公司
　郭　洁　中国电建集团北京勘测设计研究院有限公司
　陈　钢　中国电建集团华东勘测设计研究院有限公司
　李仲杰　中国电建集团西北勘测设计研究院有限公司
　张　明　清华大学
　侯瑜京　中国水利水电科学研究院
　郄春生　英语高级翻译
　梁洪丽　上海勘测设计研究院有限公司

国家能源局

Announcement of National Energy Administration of the People's Republic of China [2021] No. 1

National Energy Administration of the People's Republic of China has approved and issued 320 energy sector standards including *Code for Integrated Resettlement Design of Hydropower Projects* (Attachment 1), the foreign language version of 113 energy sector standards including *Carbon Steel and Low Alloy Steel for Pressurized Water Reactor Nuclear Power Plants—Part 7: Class 1, 2, 3 Plates* (Attachment 2), and the amendment notification for 5 energy sector standards including *Technical Code for Investigation and Assessment of Aquatic Ecosystem for Hydropower Projects* (Attachment 3).

Attachments: 1. Directory of Sector Standards

2. Directory of Foreign Language Version of Sector Standards

3. Amendment Notification for Sector Standards

National Energy Administration of the People's Republic of China

January 7, 2021

Attachment 1:

Directory of Sector Standards

Serial number	Standard No.	Title	Replaced standard No.	Adopted international standard No.	Approval date	Implementation date
...						
15	NB/T 10498-2021	Code for Design of AC 110 kV ~ 500 kV Power Cable Systems for Hydropower Station	DL/T 5228-2005		2021-01-07	2021-07-01
...						

Foreword

According to the requirements of Document GNKJ [2016] No. 238 issued by National Energy Administration of the People's Republic of China, "Notice on Releasing the Development and Revision Plan of Energy Sector Standards in 2016", and after extensive investigation and research, summarization of practical experiences, consultation of relevant international standards, and wide solicitation of opinions, the drafting group has prepared this code.

The main technical contents of this code include: general provisions, terms and symbols, service conditions, selection of main technical parameters, type and structure, terminations and joints, earthing methods and overvoltage protection for metallic sheath, auxiliary facilities, cable layout and laying, fire protection, and test.

The main technical contents revised are as follows:

—Adding the requirements for cable on-line monitoring device.

—Adding the requirements for cable fire prevention.

—Adding the chapter "Test".

—Incorporating the content of voltage, current-carrying capacity and cross-sectional area of conductors, rated short-time withstand current and insulation level into the chapter "Selection of Main Technical Parameters".

—Deleting the content of equalizing cable and oil-filled cable.

National Energy Administration of the People's Republic of China is in charge of the administration of this code. China Renewable Energy Engineering Institute has proposed this code and is responsible for its routine management. Energy Sector Standardization Technical Committee on Hydropower Electrical Design (NEA/TC17) is responsible for the explanation of specific technical contents. Comments and suggestions in the implementation of this code should be addressed to:

China Renewable Energy Engineering Institute
No. 2 Beixiaojie, Liupukang, Xicheng District, Beijing 100120, China

Chief development organization:

POWERCHINA Chengdu Engineering Corporation Limited

Participating development organizations:

POWERCHINA Huadong Engineering Corporation Limited

Jiangsu Ankura Smart Transmission Engineering Technology Co., Ltd.

Chief drafting staff:

LI Qiang	HE Xiao	WANG Yaohui	WEN Fengxiang
YU Dan	YU Peng	LI Yong	FENG Zhenqiu
HOU Yanshuo	WANG Jingwen	YI Xiaojing	CHEN Xiaoming
WANG Xinqi	JU Lin		

Review panel members:

YU Qinggui	PANG Xiulan	WANG Jinfu	KANG Benxian
CHEN Yinqi	WU Zhongping	ZHOU Huigao	XIA Fujun
YANG Jianjun	WANG Yong	SHAO Guangming	YANG Junshuang
WANG Huajun	LIU Changwu	DENG Shuangxue	XIE Xiaohui
PAN Hong	ZONG Wanbo	YANG Mei	

Contents

1	**General Provisions**	1
2	**Terms and Symbols**	2
2.1	Terms	2
2.2	Symbols	2
3	**Service Conditions**	3
3.1	Operating Conditions	3
3.2	Laying Conditions	3
4	**Selection of Main Technical Parameters**	6
4.1	Voltage	6
4.2	Cross-Sectional Area of Conductors	6
4.3	Insulation Level	8
5	**Type and Structure**	9
5.1	Type	9
5.2	Conductor	9
5.3	Insulation	9
5.4	Conductor Screen and Insulation Screen	12
5.5	Buffer	12
5.6	Metallic Sheath	12
5.7	Outer Sheath	13
6	**Terminations and Joints**	15
6.1	General Requirements	15
6.2	Terminations	16
6.3	Joints	17
7	**Earthing Methods and Overvoltage Protection for Metallic Sheath**	19
7.1	Selection of Metallic Sheath Earthing Methods	19
7.2	Induced Voltage on Metallic Sheath	21
7.3	Shielding Conductor	22
7.4	Sheath Voltage Limiter	23
8	**Auxiliary Facilities**	25
8.1	Cable Clamp	25
8.2	Cable Support	25
8.3	Cable Termination Support	25
8.4	Cable On-Line Monitoring Device	26
9	**Cable Layout and Laying**	27
9.1	General Requirements	27

	9.2	Cable Layout ··27
	9.3	Layout of Terminations and Joints ···28
	9.4	Cable Fixing ··29
	9.5	Cable Laying ···29
10		**Fire Protection** ···30
11		**Test** ··31

Explanation of Wording in This Code··33
List of Quoted Standards··34

1 General Provisions

1.0.1 This code is formulated with a view to standardizing the design of AC 110 kV to 500 kV power cable systems for hydropower stations, to achieve the objectives of safety, reliability, technological advancement, cost effectiveness, and ease of construction and maintenance.

1.0.2 This code is applicable to the design of power cable systems with a nominal voltage of AC 110 kV to 500 kV and a frequency of 50 Hz for the construction, renovation and extension of hydropower stations.

1.0.3 In addition to this code, the design of AC 110 kV to 500 kV power cable systems for hydropower stations shall comply with other current relevant standards of China.

2 Terms and Symbols

2.1 Terms

2.1.1 shielding conductor

insulated wire or cable with both ends earthed, which is parallel to the single-conductor power cable, forcing single-phase short-circuit current to flow back to the power source, to reduce the induced overvoltage on the cable metallic sheath and protect the insulation of outer sheath

2.1.2 slip fixing

fixing method that allows cables an axial angle change at the attachment point or slight transverse displacement to accommodate thermal expansion and contraction

2.1.3 rigid fixing

fixing method by clamping the cable to prevent any displacement due to thermal expansion and contraction

2.1.4 snaking

cable laying method by which cables are arranged in a wave shape to reduce axial thermal stress or to facilitate free expansion and contraction of cables according to the design requirements

2.2 Symbols

U_0 — rated r.m.s. power-frequency voltage between each conductor and the screen or metallic sheath, for which the cables and accessories are designed

U — rated r.m.s. power-frequency voltage between any two conductors, for which the cables and accessories are designed

U_m — maximum r.m.s. power-frequency voltage between any two conductors, for which the cables and accessories are designed

U_{p1} — peak value of lightning impulse withstand voltage between each conductor and the screen or metallic sheath, for which the cables and accessories are designed

U_{p2} — peak value of switching impulse withstand voltage between each conductor and the screen or metallic sheath, for which the cables and accessories are designed

3 Service Conditions

3.1 Operating Conditions

3.1.1 The nominal system voltage and the highest system voltage shall be in accordance with Table 3.1.1.

Table 3.1.1 Nominal system voltage and highest system voltage (kV)

Nominal system voltage	110	220	330	500
Highest system voltage	126	252	363	550

3.1.2 The lightning impulse voltage and switching impulse voltage shall be determined by the system basic insulation level and insulation coordination.

3.1.3 The system frequency shall be 50 Hz.

3.1.4 The neutrals of 110 kV and above systems shall be effectively earthed.

3.1.5 The cables and accessories shall meet the requirements of the altitude and seismic intensity at the location of installation, taking into account the impacts of environmental conditions such as temperature, soil thermal resistance, wind speed, icing, and solar radiation.

3.1.6 The maximum operating current of the cable shall be determined on the basis of the maximum current value under continuous operation, emergency operation and overload operation conditions.

3.1.7 The symmetrical and asymmetrical short-circuit currents flowing through the cable in the case of phase-to-phase and phase-to-earth short circuits at the cable power-end shall be determined according to the long-term development plan of the power system.

3.1.8 The duration of rated short-time withstand current may be taken as 3 s for 110 kV power cables, and 2 s for power cables of 220 kV and above. The duration of single-phase-to-earth short-circuit current shall not be less than the action time of the first-level backup relay protection. For cables with dual main protection, the duration of the single-phase-to-earth short-circuit current may be determined by the action time of the dual main protection.

3.2 Laying Conditions

3.2.1 The cable design shall consider the following laying conditions:

  1   The routing, terrain, altitude difference and length of cable lines.

2   The number of cable circuits, and the pattern and spacing of the cables in horizontal, vertical or triangular arrangement.

3   Earthing methods of metallic sheath.

4   Arrangement of cable fire prevention installation.

5   Special laying methods and other special requirements.

3.2.2 The ambient temperature for cable laying shall be in accordance with Table 3.2.2.

Table 3.2.2   Ambient temperature for cable laying

Laying method		Ambient temperature
Underground	Direct burial	Average ground temperature of the hottest month at the burial depth
	Protection conduit	Average ground temperature of the hottest month at the burial depth
In air	Tunnel with ventilation	Design ventilation temperature
	Tunnel without ventilation or cable trench	Average of daily maximum temperatures of the hottest month plus 5 °C
	Overhead with solar radiation	Average of daily maximum temperatures of the hottest month

3.2.3 In addition to Articles 3.2.1 and 3.2.2 of this code, the following factors shall be considered for underground cables and accessories:

1   The metallic sheath structure and armor type determined by installation conditions for cables in duct banks or directly buried.

2   The outer sheath material determined by the installation conditions such as anti-corrosion, moisture-proof, and small animal prevention.

3   The burial depth of cables and the thickness of frozen soil.

4   The type of soil such as sand, clay, backfill and other soils along the cable route and their thermal resistance coefficients, indicating whether these data are measured or assumed.

5   The maximum, minimum and average temperatures of the soil at the burial depth.

6   The detailed information on nearby operating cables and/or other heat sources.

7   The length of cable trenches or duct banks, and the distance between manholes if any.

8   The number, inner diameter and materials of duct banks.

9   The distance between duct banks.

4 Selection of Main Technical Parameters

4.1 Voltage

The cable voltage shall be in accordance with Table 4.1.1.

Table 4.1.1   Cable voltage (kV)

U	U_m	U_0
110	126	64
220	252	127
330	363	190
500	550	290

4.2 Cross-Sectional Area of Conductors

4.2.1 Under the service conditions, the current-carrying capacity of the conductor cross section shall not be less than the maximum continuous operating current. The current-carrying capacity of the conductor can be calculated by the following formula:

$$I = \left[\frac{(\theta_C - \theta_0) - W_d[0.5T_1 + (T_2 + T_3 + T_4)]}{RT_1 + R(1+\lambda_1)T_2 + R(1+\lambda_1+\lambda_2)(T_3 + T_4)} \right]^{\frac{1}{2}} \qquad (4.2.1)$$

where

- I is the current-carrying capacity of the conductor (A);
- θ_C is the allowable maximum conductor temperature under continuous operation of cables (°C), taken as 90 °C for normal operation of cross-linked polyethylene (XLPE) cable conductor;
- θ_0 is the ambient temperature during continuous operation of the cable (°C);
- W_d is the dielectric loss per meter of the insulation (W/m);
- T_1 is the thermal resistance per meter of the insulation (K · m/W);
- T_2 is the thermal resistance per meter of the bedding between the metallic sheath and armor (K · m/W);
- T_3 is the thermal resistance per meter of the outer sheath (K · m/W);
- T_4 is the external thermal resistance per meter of the cable (K · m/W);

R is the AC resistance per meter of the cable at allowable maximum temperature in continuous operation (Ω/m);

λ_1 is the loss coefficient of the metallic sheath;

λ_2 is the loss coefficient of the armor.

4.2.2 The cross-sectional area of conductors shall meet the requirements of rated short-time withstand current.

4.2.3 The cross-sectional area for the cable short-circuit thermal stability can be calculated by the following formulae:

$$S \geq \frac{\sqrt{Q_d}}{C} \times 10^2 \qquad (4.2.3\text{-}1)$$

$$C = \sqrt{\frac{JQ}{k\rho_{20}\alpha} \ln \frac{1+\alpha(t_d - 20)}{1+\alpha(t_g - 20)}} \qquad (4.2.3\text{-}2)$$

where

S is the cross-sectional area for the cable short-circuit thermal stability (mm²);

Q_d is the thermal effect of short-circuit current (A² · s);

C is the thermal stability coefficient;

J is the coefficient of joules to calories conversion (J/cal), taken as 4.2 J/cal;

Q is the thermal capacity per unit volume of cable conductor [cal/(cm³ · °C)], taken as 0.59 cal/(cm³ · °C) for aluminum-core conductor and 0.81 cal/(cm³ · °C) for copper-core conductor;

k is the ratio of AC resistance to DC resistance of conductor at 20 °C;

ρ_{20} is the resistivity of cable conductor at 20 °C (Ω · cm²/cm), taken as 0.02826 × 10⁻⁴ Ω · cm²/cm for aluminum-core conductor and 0.01724 × 10⁻⁴ Ω · cm²/cm for copper-core conductor;

α is the conductor temperature coefficient of resistance at 20 °C (1/°C), taken as 0.00403/°C for aluminum-core conductor and 0.00393/°C for copper-core conductor;

t_d is the allowable maximum temperature of cable core conductor under short circuit (°C), taken as 250 °C for XLPE cable conductor under short circuit;

t_g is the allowable maximum temperature at rated load (°C).

4.2.4 The cross-sectional area of conductors shall be selected from the standard section series. The cross-sectional area should not be less than 240 mm^2 for 110 kV cables, 400 mm^2 for 220 kV cables, 630 mm^2 for 330 kV cables, and 800 mm^2 for 500 kV cables.

4.3 Insulation Level

4.3.1 The lightning impulse withstand voltages of cables, terminations and joints shall be in accordance with Table 4.3.1. The cables of 330 kV and above shall be subjected to overvoltage calculation and insulation coordination check.

Table 4.3.1 Lightning impulse withstand voltages of cables, terminations and joints (kV)

Cable voltage $U_0/U(U_m)$	64/110 (126)	127/220 (252)	190/330 (363)	290/500 (550)
Lightning impulse withstand voltage U_{p1}	550	1050	1175 / 1300	1550 / 1675

4.3.2 The switching impulse withstand voltages of cables with a rated voltage of 330 kV or above and their terminations and joints shall be in accordance with Table 4.3.2.

Table 4.3.2 Switching impulse withstand voltages of cables, terminations and joints (kV)

Cable voltage $U_0/U(U_m)$	190/330 (363)	290/500 (550)
Switching impulse withstand voltage U_{p2}	950	1175

4.3.3 The power frequency and lightning impulse withstand voltages of cable outer sheath insulation shall be in accordance with Table 4.3.3. The DC withstand voltage of cable outer sheath insulation should be 30 kV, and the duration shall not be less than 1 min.

Table 4.3.3 Power frequency and lightning impulse withstand voltages of cable outer sheath insulation (kV)

Cable voltage $U_0/U(U_m)$	Rated short-time power frequency withstand voltage (r.m.s.)	Lightning impulse withstand voltage (peak value)
64/110 (126)	25	37.5
127/220 (252)	25	47.5
190/330 (363)	25	62.5
290/500 (550)	25	72.5

5 Type and Structure

5.1 Type

5.1.1 110 kV to 500 kV cables should be of XLPE type.

5.1.2 The XLPE cable shall consist of conductor, conductor screen, insulation, insulation screen, buffer, metallic sheath, and outer sheath at least.

5.2 Conductor

5.2.1 The material of conductors should be copper.

5.2.2 The cable with a conductor cross-sectional area of 630 mm^2 or less should adopt compacted stranded circular conductor structure; the cable with a conductor cross-sectional area of 1000 mm^2 or above should adopt Milliken conductor structure; the cable with a conductor cross-sectional area of 800 mm^2 may adopt either compacted stranded circular conductor structure or Milliken conductor structure.

5.3 Insulation

5.3.1 XLPE cable insulations shall be produced by dry-type cross-bonding process, and the conductor screen, insulation and insulation screen of the cable shall be formed by co-extruding in one operation.

5.3.2 The nominal thickness of XLPE cable insulation shall be determined according to the power frequency withstand voltage and lightning impulse withstand voltage, and should not be less than the value specified in Table 5.3.2.

Table 5.3.2 Nominal thickness of XLPE cable insulation

$U_0 / U (U_m)$ (kV)	Nominal thickness of insulation t_n (mm)
64/110 (126)	16
127/220 (252)	24
190/330 (363)	27
290/500 (550)	30

5.3.3 The average thickness, minimum thickness and eccentricity of the insulation shall be in accordance with Table 5.3.3.

Table 5.3.3 Average thickness, minimum thickness and eccentricity of insulation

Item	110 kV, 220 kV	330 kV, 500 kV
Average thickness	$\geq t_n$	$\geq t_n$
Minimum thickness	$\geq 0.95\, t_n$	$\geq 0.95\, t_n$
Eccentricity	$\leq 6\,\%$	$\leq 5\,\%$

NOTES:

1. t_n is the nominal thickness of the insulation specified in Table 5.3.2.
2. The eccentricity is the ratio, in percentage, of the difference between the maximum and minimum thickness measured on the same section to the maximum thickness.

5.3.4 The limit values of micropores, impurities and protrusions in the insulation and the insulation interface shall be selected from Table 5.3.4.

Table 5.3.4 Limit values of micropores, impurities and protrusions in insulation and insulation interface

Voltage (kV)	Inspection item		Limit value (nos.)
110	Insulation	Micropores larger than 0.05 mm in diameter	0
		Micropores larger than 0.025 mm but smaller than or equal to 0.05 mm in diameter	$\leq 18/10\ cm^3$
		Non-transparent impurities larger than 0.125 mm in size	0
		Non-transparent impurities larger than 0.05 mm but smaller than or equal to 0.125 mm in size	$\leq 6/10\ cm^3$
		Semi-transparent dark brown impurities larger than 0.25 mm in size	0
	Interface between semi-conducting screen and insulation	Micropores larger than 0.05 mm in diameter	0
	Interface between conductor semi-conducting screen and insulation	Protrusions larger than 0.125 mm in size intruding into the insulation and the semi-conducting screen	0
	Interface between insulation semi-conducting screen and insulation	Protrusions larger than 0.125 mm in size intruding into the insulation and the semi-conducting screen	0

Table 5.3.4 *(continued)*

Voltage (kV)	Inspection item		Limit value (nos.)
220	Insulation	Micropores larger than 0.05 mm in diameter	0
		Micropores larger than 0.025 mm but smaller than or equal to 0.05 mm in diameter	$\leq 18/10\ cm^3$
		Non-transparent impurities larger than 0.125 mm in size	0
		Non-transparent impurities greater than 0.05 mm but less than or equal to 0.125 mm in size	$\leq 6/10\ cm^3$
		Semi-transparent dark brown impurities larger than 0.16 mm in size	0
	Interface between semi-conducting screen and insulation	Micropores larger than 0.05 mm in diameter	0
	Interface between conductor semi-conducting screen and insulation	Protrusions larger than 0.08 mm in size intruding into the insulation and the semi-conducting screen	0
	Interface between insulation semi-conducting screen and insulation	Protrusions larger than 0.08 mm in size intruding into the insulation and the semi-conducting screen	0
330 and 500	Insulation	Micropores larger than 0.02 mm in diameter	0
		Non-transparent impurities larger than 0.075 mm in diameter	0
	Interface between semi-conducting screen and insulation	Micropores larger than 0.02 mm in diameter	0
	Interface between conductor semi-conducting screen and insulation	Protrusions larger than 0.05 mm in size intruding into the insulation and the semi-conducting screen	0
	Interface between insulation semi-conducting screen and insulation	Protrusions larger than 0.05 mm in size intruding into the insulation and the semi-conducting screen	0

5.4 Conductor Screen and Insulation Screen

5.4.1 The conductor screen of 110 kV XLPE cables with a cross-sectional area less than 500 mm² shall be an extruded semi-conducting layer, and the conductor screen of other XLPE cables shall consist of semi-conducting tape and extruded semi-conducting layer.

5.4.2 The insulation screen shall be an extruded semi-conducting layer.

5.5 Buffer

5.5.1 There shall be a buffer between the insulation screen and the metallic sheath.

5.5.2 The buffer shall keep the insulation screen in good electrical contact with the metallic sheath.

5.5.3 The buffer thickness shall meet the requirements for accommodating the thermal expansion of the cable in operation.

5.6 Metallic Sheath

5.6.1 The metallic sheath of a cable shall be selected according to the asymmetric short-circuit current and the requirements for blocking radial water and withstanding mechanical tension and compression.

5.6.2 The selection of metallic sheaths should meet the following requirements:

1 In a non-corrosive environment, the metallic sheath of a cable may be of the corrugated or smooth aluminum type.

2 In an underwater or corrosive environment, the metallic sheath of a cable should be of the lead type.

5.6.3 The cross-sectional area of the cable metallic sheath shall meet the requirements of the short-circuit capacity in the event of a single-phase-to-earth fault or two simultaneous phase-to-earth faults at different locations, and the manufacturer shall provide the corresponding calculation sheet.

5.6.4 The minimum thickness of a cable lead sheath shall meet the requirements of the following formula:

$$t_{min} \geq t_n - (0.1 + 0.05 t_n) \tag{5.6.4}$$

where

t_{min} is the minimum thickness (mm);

t_n is the nominal thickness (mm).

5.6.5 The minimum thickness of a corrugated aluminum sheath shall meet the requirements of the following formula:

$$t_{min} \geq t_n - (0.1 + 0.15t_n) \tag{5.6.5}$$

where

t_{min} is the minimum thickness (mm);

t_n is the nominal thickness (mm).

5.7 Outer Sheath

5.7.1 The outer sheath of a cable shall adopt insulated polyvinyl chloride or polyethylene with a good thermal-resistance property, whose insulation level shall be in accordance with Table 4.3.3 of this code. The outer sheath of a cable under general laying conditions may adopt polyvinyl chloride, and the outer sheath of a cable which is directly buried, in conduit, or laid where the groundwater level is high or the temperature is low, should adopt polyethylene.

5.7.2 The flame retardance of the outer sheath of a cable shall not be inferior to Class C. The finished cables shall pass the flame propagation test specified in the current national standard GB/T 12666.2, *Test Method on a Single Wire or Cable Under Fire Conditions—Part 2: Horizontal Specimen Flame Test*.

5.7.3 The outer sheath of a cable shall comply with the current national standards GB/T 2952.1, *Protective Coverings for Electric Cables—Part 1: General*; GB/T 2952.2, *Protective Coverings for Electric Cables—Part 2: Protective Coverings for Cables with Metallic Sheath*; and GB/T 2952.3, *Protective Coverings for Electric Cables—Part 3: General Protective Coverings for Cables with Non-metallic Sheath*. In addition, there shall be a uniform and firm conducting layer on the surface.

5.7.4 The minimum thickness of the outer sheath of a cable shall satisfy the following formula:

$$t_{min} \geq t_n - (0.1 + 0.15t_n) \tag{5.7.4}$$

where

t_{min} is the minimum thickness (mm);

t_n is the nominal thickness (mm).

5.7.5 Where needed, the outer sheath shall be immune to termites, rodents and fungi, and the preventive additives for the outer sheath shall not be environmentally prohibited materials.

5.7.6 Where needed, the outer sheath may also consist of an armor and an extruded polyvinyl chloride or polyethylene layer. The armor is of non-magnetic steel tape or steel wire, and shall comply with the current national standards GB/T 2952.1, *Protective Coverings for Electric Cable—Part 1: General*; GB/T 2952.2, *Protective Coverings for Electric Cables—Part 2: Protective Coverings for Cables with Metallic Sheath*; and GB/T 2952.3, *Protective Coverings for Electric Cable—Part 3: General Protective Coverings for Cables with Non-metallic Sheath.*

5.7.7 The following markings shall be printed or imprinted on the entire outer sheath of a cable:

1. Manufacturer.
2. Voltage.
3. Conductor cross-sectional area and material.
4. Insulation material.
5. Year of manufacture.

6 Terminations and Joints

6.1 General Requirements

6.1.1 The voltage of cable terminations and joints may be expressed as $U_0/U(U_m)$ and shall not be lower than the cable voltage.

6.1.2 The insulation characteristics of cable terminations and joints shall meet the following requirements:

1 The various components, parts and materials of cable terminations and joints shall facilitate field installation provided that the design process requirements are met. The cable accessories shall be integrated with the cable with respect to the insulation structure.

2 The unified specific creepage distance of the external insulation of cable terminations shall be selected according to the pollution class, and the maximum phase-to-earth voltage of the system shall be taken when calculating the unified creepage distance of the external insulation of cable terminations. The unified specific creepage distance of the external insulation of cable terminations shall not be lower than the values in Table 6.1.2.

Table 6.1.2 Unified specific creepage distance of the external insulation of cable terminations

Pollution class	Unified specific creepage distance (mm/kV)
a	22
b	27.8
c	34.7
d	43.3
e	53.7

3 The altitude correction shall be conducted for the external insulation level of cable terminations installed at an altitude above 1000 m in accordance with the current national standard GB/T 311.1, *Insulation Co-ordination—Part 1: Definitions, Principles and Rules*.

4 The outer shell of joint shall be insulated against earth, and the insulation level of the insulation shell shall be consistent with that of the outer sheath of the cable. The insulation level of the insulated joint at the disconnection of the metallic screen shall not be less than 2 times

that of the outer sheath of the connecting cables.

6.1.3 The mechanical strength of cable terminations shall meet the following requirements:

1 The seismic design of cable terminations shall comply with the current national standard GB 50260, *Code for Seismic Design of Electrical Installations*.

2 The cable terminations directly connected to overhead lines shall be able to withstand a horizontal tension of 2 kN.

3 The outdoor cable terminations shall also be able to withstand wind pressure in the operating environment.

6.1.4 A protective box shall be provided for the insulation shell of the joint directly buried in earth and shall be corrosion-resistant. The box shall protect the insulation shell from external forces.

6.2 Terminations

6.2.1 The insulating fluid, when applied in the cable termination, shall be compatible with the stress cone material and harmless to the cable dielectrics; a separate monitoring device should be provided for the insulating liquid or gas, if any, in the cable termination.

6.2.2 For the cable terminations connected to gas-insulated switchgear (GIS) or oil-immersed transformer, dry-type cable terminations should be adopted.

6.2.3 The cable terminations connected to GIS shall meet the following requirements:

1 The design of the interface between the cable termination and the GIS and the supply scope shall comply with the current national standard GB/T 22381, *Cable Connections Between Gas-Insulated Metal-Enclosed Switchgear for Rated Voltages Equal to and above 72.5 kV and Fluid-Filled and Extruded Insulation Power Cables—Fluid-Filled and Dry Type Cable-Terminations*, and shall facilitate erection, maintenance and test of cable terminations.

2 An insulated flange shall be set between the outer shell of the cable termination and the enclosure of GIS. The sheath voltage limiters shall be evenly distributed along the circumference on both sides of the insulated flange. The insulation level of the insulated flange shall be consistent with the insulation level of the cable outer sheath.

3 The outdoor cable terminations shall be provided with a sun shield.

6.2.4 The cable terminations connected to the oil-immersed type transformer shall meet the following requirements:

1 An insulated flange shall be set between the outer shell of the cable termination and the enclosure of the transformer. The sheath voltage limiters shall be evenly distributed along the circumference of both sides of the insulated flange. The insulation level of the insulated flange shall be consistent with that of the cable outer sheath.

2 Measures shall be taken to prevent cable dielectrics from contacting transformer oil.

6.2.5 The cable terminations connected to overhead lines shall meet the following requirements:

1 Base insulators shall be provided to insulate the termination base from the support, whose insulation level shall be consistent with that of the cable outer sheath.

2 The unified specific creepage distance of external insulation of cable terminations shall be determined considering the environmental conditions such as pollution class and altitude.

3 The cable terminations for rated voltages of 330 kV and above shall be equipped with shield rings, and the cable terminations for rated voltage of 220 kV should be equipped with shield rings.

6.3 Joints

6.3.1 Cables should not be set with joints when there are no restrictions on transportation and laying conditions, or the selection of sheath voltage limiters.

6.3.2 The straight joints or sectionalizing joints shall be adopted according to the earthing method of metallic sheath.

6.3.3 The joints shall meet the following requirements:

1 The rated voltage and insulation level of joints shall not be lower than those of the cables.

2 The conductors of joints shall be well connected, and the crimping connection should be adopted.

3 The joints shall have an insulation shell with good sealing performance, and the protective box outside the insulation shell shall be able to withstand external mechanical forces.

6.3.4 The selection of joint structure type shall meet the following requirements:

1 The joint structure shall meet the requirements of cable voltage level, type of insulation, installation environment and equipment reliability.

2 The type of joint structure may be selected in accordance with Table 6.3.4.

Table 6.3.4 Type of joint structure

Type of cable insulation	Voltage level (kV)	Type of structure	Structure characteristics
XLPE	110 - 500	One piece premolded joint	The main components are made of rubber and premolded. The inner diameter of the premolded components and the outer diameter of cables adopt interference fit to ensure sufficient contact pressure at the interface
		Composite type prefabricated joint	Prefabricated rubber stress cones and prefabricated epoxy insulation parts are assembled on site and compressed by springs

7 Earthing Methods and Overvoltage Protection for Metallic Sheath

7.1 Selection of Metallic Sheath Earthing Methods

7.1.1 The metallic sheath of the AC single-core power cable shall be earthed solidly at least at one end. Under normal operating conditions, the maximum induced voltage at any non-earthing end shall meet the following requirements:

1 The induced voltage at any non-earthing end shall not be larger than 50 V when no effective safety protection measure is taken to prevent personnel from touching the metallic sheath. Safety protection measures shall be taken when the induced voltage exceeds 50 V.

2 Except for the above conditions, the induced voltage at any non-earthing end shall not be larger than 300 V.

7.1.2 The single-point direct earthing shall meet the following requirements:

1 The single-point direct earthing shall be adopted (Figure 7.1.2-1) where cable lines are short and meet the requirements for the induced voltages specified in Article 7.1.1 of this code. The earthing point shall be set at one end of the line, and the other end is earthed via sheath voltage limiters.

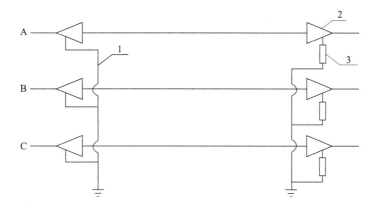

Key

1 direct earthing

2 cable termination

3 sheath voltage limiter

Figure 7.1.2-1   Single-point direct earthing

2 The mid-point direct earthing shall be adopted (Figure 7.1.2-2) where cable lines are long and the single-point direct earthing at one end cannot satisfy the requirements for induced voltage specified in Article

7.1.1 of this code. The earthing point shall be set in the middle of the line, and both ends are earthed via sheath voltage limiters.

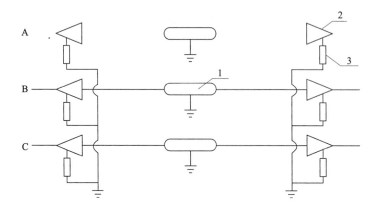

Key

1 joint
2 cable termination
3 sheath voltage limiter

Figure 7.1.2-2 Mid-point direct earthing

3 The earthing point of the cable metallic sheath may be selected at either end of the cable. When two ends of the cable are connected with different electrical equipment, the earthing point shall be selected as follows:

1) When one end of the cable is connected to the transformer and the other is connected to the overhead line, the earthing point of the metallic sheath should be set at the end connected to the overhead line, with cross-bonding method.

2) When one end of the cable is connected to the GIS and the other is connected to the overhead line, the earthing point of the metallic sheath shall be set at the end connected to the overhead line, with cross-bonding method.

3) When one end of the cable is connected to the GIS and the other is connected to the transformer, the earthing point of the metallic sheath should be set at the end connected to the GIS, with cross-bonding method.

7.1.3 The cross-bonding earthing (Figure 7.1.3) shall be adopted when cable lines are long and single point direct earthing cannot satisfy the specified requirements for induced voltage. The total length of cable lines should be equally divided into several sections which are multiples of 3, and the section length difference of a connection unit should not exceed 50 m. The metallic sheath of two adjacent sections shall be cross-bonded, and the metallic sheath shall adopt cross-bonding earthing at both ends of the cable line. The sectionalizing joints shall be adopted for separation of metallic sheath in the same unit. The section length shall be such that the induced voltage at any point of the metallic sheath meets the requirements of Article 7.1.1 of this code.

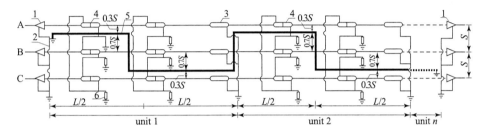

Key

1 cable termination

2 direct earthing

3 straight joint

4 sectionalizing joint

5 shielding conductor

6 sheath voltage limiter

Figure 7.1.3 Cross-bonding earthing

7.2 Induced Voltage on Metallic Sheath

7.2.1 The calculation of the induced voltage of the metallic sheath shall comply with the current national standard GB 50217, *Standard for Design of Cables of Electric Power Engineering*.

7.2.2 The shielding conductor should be arranged in the seventy-thirty ratio when calculating the power frequency overvoltage of the cable metallic sheath.

7.2.3 The metallic sheath induced voltage may be calculated for only one section when the metallic sheath is cross-bonded and earthed directly at both ends.

7.2.4 Sheath voltage limiters shall be installed at the end of the cable metallic sheath that is not earthed directly.

7.2.5 The lightning impulse overvoltage on metallic sheath insulation can be calculated by the following formula:

$$U_\mathrm{F} = U_\mathrm{b} + I_\mathrm{m}\left(\frac{1}{2}R_\mathrm{c} + R_L L\right) \tag{7.2.5}$$

where

- U_F is the lightning impulse overvoltage on metallic sheath insulation (kV);
- U_b is the residual voltage of sheath voltage limiter (kV);
- I_m is the maximum amplitude of lightning current passing through the sheath voltage limiter (kA);
- R_c is the impulse earthing resistance (Ω) between the sheath voltage limiter and the earthing network, which is usually taken as 2 Ω;
- R_L is the equivalent resistance (Ω/m) of the impulse inductance of the connecting wire of the sheath voltage limiter, which may be taken as 0.337 Ω/m for 110 kV to 220 kV cables, 0.62 Ω/m for 330 kV cables, and 0.818 Ω/m for 500 kV cables;
- L is the length of the connecting wire of the sheath voltage limiter (m), which may be taken as 2.5 m.

7.2.6 The lightning impulse overvoltage and the switching impulse overvoltage on metallic sheath insulation should be calculated by computer simulation.

7.3 Shielding Conductor

7.3.1 When the metallic sheath of 110 kV cable and above adopts direct earthing at one end, a shielding conductor shall be installed adjacent to the cable in any of the following cases:

1. The power frequency induced voltage on the metallic sheath of cable caused by system short circuit exceeds the insulation withstand strength of the cable sheath or the power frequency withstand voltage of the sheath voltage limiter.

2. The electrical interference of cables to the adjacent weak current circuits needs to be suppressed.

7.3.2 The selection and configuration of shielding conductor shall meet the following requirements:

1. The impedance of the shielding conductor and the earthing resistance at both ends shall be able to suppress the power frequency induced overvoltage of metallic sheath of cable, and the cross-sectional area of the shielding conductor shall satisfy the thermal stability requirements under the maximum asymmetric short-circuit current. The duration of short-time withstand current shall be in accordance with Article 3.1.8

of this code.

2　The shielding conductor should be arranged in the seventy-thirty ratio. In the case of difficulty in arrangement and installation, the layout of symmetrical points on the hyperbola may be adopted.

3　Copper-core cables should be used for the shielding conductor, and their insulation level shall match the insulation of cable metallic sheath based on the calculated induced voltage. 10 kV cables should be adopted.

7.4　Sheath Voltage Limiter

7.4.1　The sheath voltage limiter shall meet the following requirements:

1　Under the possible maximum impulse current, the coordination coefficient between the impulse withstand voltage of cable outer sheath and the residual voltage of sheath voltage limiters shall not be less than 1.4, and the insulation margin of the impulse voltage on the metallic sheath and the impulse withstand voltage level on the outer sheath shall be verified.

2　Under any conditions, the power frequency overvoltage of metallic sheath shall be lower than the power frequency withstand voltage of the outer sheath insulation, and the safety factor may be taken as 1.2. Sheath voltage limiters shall be able to withstand the maximum power frequency overvoltage for the possible fault duration, which shall be taken as 2 s.

3　Sheath voltage limiters shall not be damaged after 20 possible maximum impulse current impacts.

4　The ratio of impulse residual voltage to power frequency withstand voltage for the sheath voltage limiter should be 2.0 to 3.0.

5　The sheath voltage limiters shall be oxide ones equipped with action recorders.

7.4.2　The performance parameters of the sheath voltage limiter shall meet the requirements of insulation coordination.

7.4.3　The configuration and connection of sheath voltage limiters shall meet the following requirements:

1　The configuration of the sheath voltage limiter shall be determined comprehensively based on such factors as the suppression effect of impulse overvoltage, matching of parameters under the power frequency induced overvoltage and ease of inspection and maintenance. The metallic sheath, which is directly earthed at one end and connected to the three-phase sheath voltage limiter at the other end in Y_0 pattern, should be adopted.

2 The connection circuit of sheath voltage limiters shall meet the following requirements:

1) The connecting wires shall be as short as possible, and a coaxial cable should be adopted. The cross section of the cable shall meet the requirements for the system thermal stability in the case of the maximum asymmetric short-circuit current.

2) The insulation performance of the insulated conductors connecting the circuits shall not be less than the insulation level of the outer sheath of the cable.

3) The material and IP rating of the direct earthing box of metallic sheath and the earthing box of sheath voltage limiter shall meet the requirements of their operating environment.

8 Auxiliary Facilities

8.1 Cable Clamp

8.1.1 The clamps for fixing cables should be made of aluminum alloy, or may be made of non-magnetic stainless steel, and shall have a mechanical strength to withstand the action of the short-circuit electrodynamic force. The clamps shall have smooth surface without sharp corners and burs, and be easy to install.

8.1.2 An insulating cushion made of neoprene or other synthetic materials shall be set between the clamp and the cable, and should have a thickness no less than 5 mm.

8.1.3 The clamp shall be tightened with a torque wrench.

8.1.4 The mechanical strength of the nylon belt and clamp used for bundling up and suspending cables shall meet the load requirements of cable suspension.

8.2 Cable Support

8.2.1 The mechanical strength of cable supports, slings for suspending cables and cantilever support beams shall be sufficient for carrying the total load of cables and their fixings and the temporary additional loads during installation and maintenance. The temporary additional loads during installation and maintenance shall mainly include axial tensile force, weight of cable transmission device and human body weight.

8.2.2 Cable supports shall have smooth surface without sharp corners and burs.

8.2.3 Cable supports shall be protected against corrosion and reliably earthed.

8.2.4 Cable supports shall meet the fire and corrosion protection requirements, and should be made of hot-dip galvanized profile steel. When a closed magnetic circuit is formed, non-magnetic materials shall be used to cut off the closed magnetic circuit.

8.2.5 When the rigid fixing cable supports are fixed on the structure, neither the cable nor the supports shall displace.

8.2.6 When the slip fixing cable supports are fixed on the structure, the cable supports shall not displace, and the cable may be offset axially and radially.

8.3 Cable Termination Support

8.3.1 The design of the height and spacing of cable termination supports shall meet the following requirements:

1 The mounting height of the outdoor cable termination base perpendicular to the ground shall not be less than 2.5 m.

2 The design of the cable termination supports shall meet the requirements of the cable bending radius and facilitate termination installation.

8.3.2 The cable termination support shall be able to carry the total load of the termination and the temporary additional loads during installation and maintenance, and shall comply with Article 6.1.3 of this code; the design of steel structure components shall comply with the current national standard GB 50017, *Standard for Design of Steel Structures*.

8.3.3 The material and structure of cable termination supports shall meet the following requirements:

1 The cable termination supports shall meet the anti-corrosion requirements and should be made of hot-dip galvanized profile steel.

2 The steel support should not form a closed magnetic circuit around the cable; otherwise, non-magnetic materials shall be used to cut off the closed magnetic circuit.

3 The cable termination supports shall be reliably connected to the earthing network.

8.4 Cable On-Line Monitoring Device

8.4.1 The cables of 220 kV and above should be equipped with distributed optical fiber temperature measurement on-line monitoring devices.

8.4.2 The 500 kV cable terminations and joints may be equipped with partial discharge monitoring devices.

8.4.3 The multi-point earthing on-line monitoring device should be installed at the earthing box.

8.4.4 When there is insulating oil or gas in the air termination, a separate monitoring device should be provided.

9 Cable Layout and Laying

9.1 General Requirements

9.1.1 The selection of cable routes shall meet the following requirements:

1 Avoid cable damages caused by external mechanical force, vibration, water immersion, corrosion, etc.

2 Facilitate cable laying and maintenance.

3 Avoid flammable and explosive places.

4 A certain length should be reserved at each end of the cable for making a new termination.

9.1.2 The cable route shall meet the requirements for the permissible bending radius of the cable. The permissible bending radius may be taken as 20 times the outer diameter of the cable during cable laying, and may be taken as 15 times the outer diameter of the cable after cable installation.

9.1.3 The cables and heating pipeline shall not be arranged in the same structure.

9.1.4 Cables in the project area may be laid in tunnels, vertical shafts or inclined shafts according to the engineering conditions. When the conditions are limited, they may also be laid in cable trenches, but direct burial should not be adopted.

9.1.5 The cables shall be laid in conduits or tunnels when crossing roads or railways. The inner diameter of the conduit should not be less than 1.5 times the outer diameter of the cable and the length shall extend 2 m at each end of the width of the road or railway. The distance from the conduit top to the road surface or railway subgrade shall not be less than 1 m.

9.1.6 For single-core power cables that are not arranged in trefoil, the mutual influence between phases or circuits shall be considered in the cable design when two or more circuits are arranged in the same raceway.

9.1.7 A test area shall be reserved.

9.2 Cable Layout

9.2.1 The number of cable circuits laid in the same tunnel, inclined shaft or vertical shaft should not exceed 4.

9.2.2 The clear height of cable tunnels and inclined shafts shall not be less than 2000 mm. The clear width of the passage shall not be less than 1000 mm when cables are arranged on both sides, and shall not be less than 900 mm when cables are arranged on one side.

9.2.3 The distance between adjacent layers of 110 kV to 220 kV cable brackets or hangers should not be less than 350 mm. The distance between adjacent layers of 330 kV and above cable brackets or hangers should not be less than 400 mm. The clear distance from the top cable bracket to the roof or beam bottom of the building should not be less than 500 mm when laying cables horizontally, and the distance from the bottom cable bracket to the floor shall not be less than 150 mm.

9.2.4 The vertical shaft shall be provided with space for access. If the height difference exceeds 5 m, staircases should be provided and stair landings should be provided at an interval of 3 m. If the height difference exceeds 20 m and the number of cable circuits is 2 or more, an elevator may be provided.

9.2.5 For cable tunnels, vertical shafts and inclined shafts, measures shall be taken to prevent water leakage and external water ingress, and drainage facilities shall be provided. Side drains should be provided at the bottom.

9.2.6 The inclination of the cable inclined shaft should not exceed 35°.

9.3 Layout of Terminations and Joints

9.3.1 The layout of cable terminations shall meet the following requirements:

1. The cable termination supports shall be able to facilitate the pulling of cables and the hoisting of the cable termination and its accessories.

2. The base insulator shall be so designed that the insulator can be replaced without hoisting the termination.

3. The metallic sheath earthing connection box should be arranged on the bracket, and the sheath voltage limiter shall be arranged in the earthing connection box or on the bracket beyond reach. The layout shall ensure the shortest connection of coaxial cables.

9.3.2 The layout of joints shall meet the following requirements:

1. The joints should be set in horizontal route sections.

2. The joints shall be staggered for three phases, and the layout of joints shall not hinder the fabrication of the joints and shall facilitate the joint hoisting in place.

3. The metallic sheath earthing connection box should be arranged on the bracket, and the sheath voltage limiter shall be arranged in the earthing connection box or the bracket beyond reach. The layout shall ensure

the shortest connection of coaxial cables.

9.4 Cable Fixing

9.4.1 When cables are laid in air in structures, the fixing of cables shall meet the following requirements:

1. At least 2 rigid fixings shall be arranged for cables adjacent to terminations, joints or turns.

2. At least 2 rigid fixings should be arranged on the top of cables laid in a vertical or inclined shaft.

3. At least 1 rigid fixing shall be arranged in each wave cycle in the snaking of cables, and the slip fixing should be used for the rest. Rigid fixings should be used at the transition from snaking to straight routing.

9.4.2 When laying cables in a vertical shaft, measures shall be taken to prevent the relative displacement between the cable conductor and the metallic sheath.

9.5 Cable Laying

9.5.1 The spacing between the brackets, hangers and/or hooks for horizontally laid cables should not be greater than 2000 mm. The spacing between the brackets, hangers and/or hooks for cables laid in vertical shafts should not be greater than 3000 mm.

9.5.2 For the snaking cables with slip fixings, the selection of parameters for snaking shall ensure that the cable axial thermal stress due to temperature change is non-destructive to the cable insulation and does not cause fatigue fracture of cable metallic sheath in long-term operation. The snaking pitch and the initial snaking amplitude shall be calculated by the manufacturer according to the engineering conditions.

10 Fire Protection

10.0.1 When 110 kV to 500 kV cables and power or control cables are laid in the same cable raceway, gallery or trench, fire barriers shall be installed between 110 kV to 500 kV cables and power or control cables.

10.0.2 When cables pass through floor slabs, floors, partitions, and upper and lower ends of vertical shafts, the entrances and exits, branches and cable holes of cable raceways, galleries or trenches shall be plugged with noncombustible materials with a fire resistance rating of not less than 1 h.

10.0.3 The layout of cable fire partitions shall meet the following requirements:

1. Fire partitions should be set at the entrance, exit and branch of cable raceways, galleries or trenches, and at an interval of 60 m in long cable tunnels and trenches.

2. The firewall shall be made of noncombustible materials with a fire resistance rating of not less than 1 h.

3. The doors in the fire partition shall have a fire resistance rating no less than 0.5 h. In the case of no fire doors, measures shall be taken to prevent cross-fire within a cable section of 1 m on each side of the fire partition.

10.0.4 The fireproof plugging of a vertical shaft shall meet the following requirements:

1. The fireproof plugging shall be set at the top and bottom ends, entrance and exit, and each floor of the vertical shaft.

2. When two or more cable circuits are laid in the same shaft, fire barriers shall be used to separate the circuits.

3. When water spray and water mist fire extinguishers are provided in the shaft, the fireproof plugging in the shaft need not be subject to Items 1 and 2 of this Article.

4. The vertical shaft shall be plugged with noncombustible materials with a fire resistance rating of not less than 1 h. The plugged zone shall be able to bear the load of the inspectors. The manhole may be covered by a load-bearing fire barrier.

10.0.5 The cable brackets shall not be made of flammable materials, and the bracket design shall consider the installation of fire barriers.

11 Test

11.0.1 The tests of cables and their accessories shall comply with the current national standards GB/T 11017.1, *Power Cables with Cross-Linked Polyethylene Insulation and Their Accessories for Rated Voltage of 110 kV (U_m = 126 kV)—Part 1: Test Methods and Requirements*; GB/T 11017.2, *Power Cables with Cross-Linked Polyethylene Insulation and Their Accessories for Rated Voltage of 110 kV (U_m = 126 kV)—Part 2: Power Cables*; GB/T 11017.3, *Power Cables with Cross-Linked Polyethylene Insulation and Their Accessories for Rated Voltage of 110 kV (U_m = 126 kV)—Part 3: Accessories*; GB/T 18890.1, *Power Cables with Cross-Linked Polyethylene Insulation and Their Accessories for Rated Voltage of 220 kV (U_m = 252 kV)—Part 1: Test Methods and Requirements*; GB/T 18890.2, *Power Cables with Cross-Linked Polyethylene Insulation and Their Accessories for Rated Voltage of 220 kV (U_m = 252 kV)—Part 2: Power Cables*; GB/T 18890.3, *Power Cables with Cross-Linked Polyethylene Insulation and Their Accessories for Rated Voltage of 220 kV (U_m = 252 kV)—Part 3: Accessories*; GB/T 22078.1, *Power Cables with Cross-Linked Polyethylene Insulation and Their Accessories for Rated Voltage of 500 kV (U_m = 550 kV)— Part 1: Power Cable Systems-Cables with Cross-Linked Polyethylene Insulation and Their Accessories for Rated Voltage of 500 kV (U_m = 550 kV)—Test Methods and Requirements*; GB/T 22078.2, *Power Cables with Cross-Linked Polyethylene Insulation and Their Accessories for Rated Voltage of 500 kV (U_m = 550 kV)—Part 2: Power Cables with Cross-Linked Polyethylene Insulation for Rated Voltage of 500 kV (U_m = 550 kV)*; and GB/T 22078.3, *Power Cables with Cross-Linked Polyethylene Insulation and Their Accessories for Rated Voltage of 500 kV (U_m = 550 kV)—Part 3: Accessories for Power Cables with Cross-Linked Polyethylene Insulation for Rated Voltage of 500 kV (U_m = 550 kV)*.

11.0.2 The cables of 110 kV and above shall be subjected to type tests, and the cables of 220 kV and above shall be subjected to pre-qualification tests.

11.0.3 The field AC voltage withstand test for 500 kV cables shall meet the following requirements:

1 For the field insulation AC voltage test of 500 kV cables, the voltage waveform shall be approximately sinusoidal, and the frequency shall be 20 Hz to 300 Hz.

2 Based on actual test conditions, when an AC voltage of 320 kV or 493

kV ($1.7U_0$) is applied to 500 kV cables, the test duration shall not be less than 1 h; when an AC voltage of 290 kV (U_0) is applied, the test duration shall not be less than 24 h; for 500 kV cable insulation test, an AC voltage of $1.7U_0$ should be applied for 1 h.

Explanation of Wording in This Code

1. Words used for different degrees of strictness are explained as follows in order to mark the differences in executing the requirements in this code.

    1) Words denoting a very strict or mandatory requirement:

    "Must" is used for affirmation, "must not" for negation.

    2) Words denoting a strict requirement under normal conditions:

    "Shall" is used for affirmation, "shall not" for negation.

    3) Words denoting a permission of a slight choice or an indication of the most suitable choice when conditions permit:

    "Should" is used for affirmation, "should not" for negation.

    4) "May" is used to express the option available, sometimes with the conditional permit.

2. "Shall meet the requirements of…" or "shall comply with…" is used in this code to indicate that it is necessary to comply with the requirements stipulated in other relative standards and codes.

List of Quoted Standards

GB 50017,	Standard for Design of Steel Structures
GB 50217,	Standard for Design of Cables of Electric Power Engineering
GB 50260,	Code for Seismic Design of Electrical Installations
GB/T 311.1,	Insulation Co-ordination—Part 1: Definitions, Principles and Rules
GB/T 2952.1,	Protective Coverings for Electric Cables—Part 1: General
GB/T 2952.2,	Protective Coverings for Electric Cables—Part 2: Protective Coverings for Cables with Metallic Sheath
GB/T 2952.3,	Protective Coverings for Electric Cables—Part 3: General Protective Coverings for Cables with Non-metallic Sheath
GB/T 11017.1,	Power Cables with Cross-Linked Polyethylene Insulation and Their Accessories for Rated Voltage of 110 kV (U_m = 126 kV)—Part 1: Test Methods and Requirements
GB/T 11017.2,	Power Cables with Cross-Linked Polyethylene Insulation and Their Accessories for Rated Voltage of 110 kV (U_m = 126 kV)—Part 2: Power Cables
GB/T 11017.3,	Power Cables with Cross-Linked Polyethylene Insulation and Their Accessories for Rated Voltage of 110 kV (U_m = 126 kV)—Part 3: Accessories
GB/T 12666.2,	Test Method on a Single Wire or Cable Under Fire Conditions—Part 2: Horizontal Specimen Flame Test
GB/T 18890.1,	Power Cables with Cross-Linked Polyethylene Insulation and Their Accessories for Rated Voltage of 220 kV (U_m = 252 kV)—Part 1: Test Methods and Requirements
GB/T 18890.2,	Power Cables with Cross-Linked Polyethylene Insulation and Their Accessories for Rated Voltage of 220 kV (U_m = 252 kV)—Part 2: Power Cables
GB/T 18890.3,	Power Cables with Cross-Linked Polyethylene Insulation

and Their Accessories for Rated Voltage of 220 kV (U_m = 252 kV)—Part 3: Accessories

GB/T 22078.1, *Power Cables with Cross-Linked Polyethylene Insulation and Their Accessories for Rated Voltage of 500 kV (U_m = 550 kV)—Part 1: Power Cable Systems-Cables with Cross-Linked Polyethylene Insulation and Their Accessories for Rated Voltage of 500 kV (U_m = 550 kV)—Test Methods and Requirements*

GB/T 22078.2, *Power Cables with Cross-Linked Polyethylene Insulation and Their Accessories for Rated Voltage of 500 kV (U_m = 550 kV)—Part 2: Power Cables with Cross-Linked Polyethylene Insulation for Rated Voltage of 500 kV (U_m = 550 kV)*

GB/T 22078.3, *Power Cables with Cross-Linked Polyethylene Insulation and Their Accessories for Rated Voltage of 500 kV (U_m = 550 kV)—Part 3: Accessories for Power Cables with Cross-Linked Polyethylene Insulation for Rated Voltage of 500 kV (U_m = 550 kV)*

GB/T 22381, *Cable Connections Between Gas-Insulated Metal-Enclosed Switchgear for Rated Voltages Equal to and above 72.5 kV and Fluid-Filled and Extruded Insulation Power Cables—Fluid-Filled and Dry Type Cable-Terminations*